Jeff E

199 Best Quotes from the Great Entrepreneur – Amazon, Blue Origin, Space Colonization, Leadership Principles, Failure and Success

(Powerful Lessons from the Extraordinary People Book 2)

edited by
Olivia Longray

ISBN: 9798612217763

“Work hard, have fun, and make history.”

— Jeff Bezos

Contents

Introduction

He has been called a super human, a genius, and a business icon. Jeff Bezos, the founder and chief executive of Amazon.com, Inc., truly is an unusual person. And he is certainly one of the most influential people in the world.

In the early 1990s, Amazon was a small online bookstore. Jeff Bezos resigned from a promising and highly paid position in a quantitative hedge fund in New York to start his new business.

He invested 10 thousand dollars of his own money in his startup. His parents later invested another 100 thousand dollars, although he warned them that they may never see the money again. In reality, they were not betting so much on the online store, but rather on Jeff himself and his huge potential. They were not wrong to do so.

Now Amazon is the world's largest marketplace for selling goods. The internet giant also controls a third of the video streaming market in the United States and half of the global cloud services market. Both government agencies and private companies store their data on Amazon's servers. Among them are major players such as Apple, General Electric, Unilever, Netflix, and even the CIA.

Jeff Bezos received insights on technological innovations and improving customer service at the age of 16, when he worked at McDonald's. From a very young age, he has been characterized by his intelligence and ingenuity. His mother noticed his genius and found him a school for gifted children 20 miles from their home. They had to leave at 7 o'clock every morning. The curriculum was intense and unconventional and the teachers encouraged creative thinking and an entrepreneurial spirit.

Jeff Bezos developed these qualities to the utmost. Now he is the richest person on Earth. After turning 50 years old, Jeff's fortune surpassed $100 billion. According to analysts, he makes over $8 million per hour. That means that every minute Jeff Bezos becomes 133 thousand dollars richer.

But wealth is not an end in itself for him. Jeff Bezos invests the money that Amazon brings him in projects aimed to benefit all mankind. His real passion has become his aerospace company, Blue Origin, which develops engines, launch vehicles, and spacecraft.

Jeff Bezos is sure that sooner or later the Earth will exhaust its resources and people will be forced to leave it and live in space on orbital stations. Jeff Bezos does not rule out this possibility for himself either. He continues to train hard to be physically prepared for a long-term space journey.

Part 1: LIFE LESSONS

Childhood & Upbringing

1. "I am very, very lucky. I'm lucky in so many ways. I won a lot of lotteries in life. I'm not just talking about Amazon, a certain financial lottery, for sure. I have won so many lotteries. My parents are both amazing role models. My grandfather. My mom had me when she was 17, and she says, 'Don't do that'."

—Business Insider, December 13, 2014

2. "You don't choose your passions, your passions choose you. I've been passionate about rockets and space since I was a five-year-old boy."

—theDECALIFE, January 12, 2019

3. "My grandfather taught me that it is harder to be kind than it is to be clever. That has always stuck with me."

—Esquire, January, 2002

4. “[Grandfather] was super important for me.... He had a ranch in South Texas, and I would spend my summers there, from age 4 to 16.... And one of the things that’s so interesting about that lifestyle and about my grandfather is he did everything himself. He didn’t call a vet if one of the animals was sick; he figured out what to do himself.”

—Business Insider, April 28, 2018

5. “When you’re a little kid, you have no idea how much your parents love you.”

—Esquire, January, 2002

6. “When I was 5 years old, I watched Neil Armstrong step onto the moon. It made me passionate about science, physics, math, exploration.”

—Wired, November 13, 2011

Education

7. “I have always been academically smart.”

—The Atlantic, November, 2019

8. “Cleverness is a gift, kindness is a choice. Gifts are easy – they’re given after all. Choices can be hard.”

—Entrepreneur Magazine, May 24, 2017

9. “I went to Princeton specifically to study physics.... I was doing well, but it was so much work for me, it was hard – and there were half a dozen people in my class who were so gifted, and their brains were just wired in a different way, and the things that I worked so hard to do came so effortlessly to them.”

—The Guardian, February 11, 2001

10. “If I ever have a different opinion than Warren Buffett, my regard for him is so high, I always assume I’m missing something.”

—Fortune, March 24, 2016

Growth Mode

11. “We can’t be in survival mode. We have to be in growth mode.”

—YourStory, January, 12, 2018

12. “What’s dangerous is not to evolve.”

—theDECALIFE, January 12, 2019

13. “[H]igh standards are teachable. In fact, people are pretty good at learning high standards simply through exposure. High standards are contagious. Bring a new person onto a high standards team, and they’ll quickly adapt. The opposite is also true. If low standards prevail, those too will quickly spread.”

—Jeff Bezos’s 2017 Letter to Amazon Shareholders, April 18, 2018

14. “Mediocre theoretical physicists make no progress. They spend all their time understanding other people’s progress.”

—Medium, July 4, 2013

Money & Debts

15. “I have the great luxury of my resources.... I’m not going to work on anything that I don’t think is improving civilization. I think The Washington Post does that, I think Amazon does that, and I think Blue Origin does that. In the long run, Blue is the most important.”

—Wired, October 15, 2018

16. “You’ll be competing against those who are passionate.... You have to be a missionary, not a mercenary. And, paradoxically, the missionaries end up making more money.”

—re:MARS conference in Las Vegas, June 6, 2019

17. “At the end of the day, when you're eighty years old and looking back on your life, you want to have minimized the number of regrets you have. That's what should drive people. Not how much money they have. It's regrets that I think haunt people at the end of their life.”

—Esquire, January, 2002

18. “[I believe in] conserving money for things that matter.”

—Fast Company, August 1, 2004

19. "I've made billions of dollars of failures at Amazon.com. Literally billions of dollars of failures. You might remember Pets.com or Kosmo.com. It was like getting a root canal with no anesthesia. None of those things are fun. But they also don't matter.

What really matters is, companies that don't continue to experiment, companies that don't embrace failure, they eventually get in a desperate position where the only thing they can do is a Hail Mary bet at the very end of their corporate existence."

—Business Insider, December 13, 2014

20. "Debt is a useful invention. It's why people can afford houses before they reach seventy. They can actually have the house while they need it and then work and pay for it. That's very, very useful."

—Esquire, January, 2002

21. "The one thing that offends me the most is when I walk by a bank and see ads trying to convince people to take out second mortgages on their home so they can go on vacation. That's approaching evil."

—Business Insider, July 16, 2015

Work-Life Balance

22. "I dance into the office every morning."

—Fortune, March 24, 2016

23. "I get asked about work-life balance all the time.... That's a debilitating phrase because it implies there's a strict trade-off. And the reality is, if I am happy at home, I come into the office with tremendous energy. And if I am happy at work, I come home with tremendous energy. It actually is a circle; it's not a balance."

—Business Insider, April 28, 2018

24. "I like to be in the office. I feel disconnected from the office if I'm traveling a lot. I travel less than 20% of the time, maybe less than 10% of the time."

—Business Insider, December 13, 2014

25. "I am lucky. I have very good relationship with [my kids]. This work-life harmony thing is what I try to teach young employees and actually senior executives at Amazon too. But especially the people coming in."

—Business Insider, April 28, 2018

Family & Kids

26. “I do the dishes every night. I’m pretty convinced it’s the sexiest thing I do.”

—Vox, December 3, 2014

27. “I’m certainly the kind of person that has to grow on a woman. It takes repeated exposure to wear down her defenses.”

—Esquire, January, 2002

28. “[My son] was the last person in his class to get a smartphone. He reminded me of this frequently. When the second-to-last person got a smartphone, he sent an email message to all of his classmates that said, ‘Then there was one’.”

—Business Insider, December 13, 2014

29. “It's impossible to interact with an eighteen-month-old child and not come away with the impression that people are fundamentally good.”

—Esquire, January, 2002

30. “Everyone always says that parenting is not a popularity contest. I think that grandparenting is.”

—Esquire, January, 2002

Happiness

31. “People think of liberty or freedom as being happiness, but it’s not. Those very smart people who wrote ‘Life, liberty, and the pursuit of happiness’ had it right.”

—Esquire, January, 2002

32. “Morale comes not from things that you layer on to make people happy. It comes from being able to build. People like to build. The question is, ‘Could it be done?’ The answer is, ‘Absolutely’.”

—Esquire, January, 2002

33. “When you’re young, deferring gratification is not a honed skill. As you get older, you get better at the marathon mentality.”

—Esquire, January, 2002

34. “People need to think of themselves as fortunate.”

—Esquire, January, 2002

At the End of Life

35. "When you are 80 years old, and in a quiet moment of reflection narrating for only yourself the most personal version of your life story, the telling that will be most compact and meaningful will be the series of choices you have made. In the end, we are our choices. Build yourself a great story."

—Jeff Bezos's commencement speech at Princeton University, May 30, 2010

36. "When I'm 80, am I going to regret leaving Wall Street? No. Will I regret missing a chance to be there at the beginning of the Internet? Yes."

—Wired, March 1, 1999

Part 2: INCREDIBLE PROJECTS

Amazon

37. "If you go back in time 18 years, I was driving the packages to the post office myself, and we were very primitive."

—CBS News, December 1, 2013

38. "The first initial start-up capital for Amazon.com came primarily from my parents, and they invested a large fraction of their life savings in what became Amazon.com.... That was a very bold and trusting thing for them to do because they didn't know. My dad's first question was, 'What's the Internet?'

So he wasn't making a bet on this company or this concept. He was making a bet on his son, as was my mother. I told them that I thought there was a 70 percent chance that they would lose their whole investment, which was a few hundred thousand dollars, and they did it anyway."

—The Academy of Achievement, May 4, 2001

39. "I knew that if I failed I wouldn't regret that, but I knew the one thing I might regret is not trying."

—theDECALIFE, January 12, 2019

40. “What I would really like people to say about Amazon is that we raised the bar on customer experience for every industry all over the world. Some companies have missions that are even bigger than the company; an example of that would be Sony.

Sony, coming out of World War II, said that their mission was, ‘We are going to make Japan known for quality.’ They had a mission that was bigger than Sony. It was a mission for Japan. And we have a similar mission.”

—Esquire, January, 2002

41. “I have a lot of passions and interests but one of them is at Amazon, the rate of change is so high and I love that. I love the pace of change. I love the fact that I get to work with these big, smart teams. The people I work with are so smart and they’re self-selected for loving to invent on behalf of customers.”

—Product Hunt, May 5, 2019

42. “The three most important things in retail are location, location, location. The three most important things for our consumer business are technology, technology, technology.”

—Inc., February 15, 2019

43. "Sometimes we measure things and see that in the short term they actually hurt sales, and we do it anyway."

—Fast Company, August 1, 2004

44. "In our retail business, we know that customers want low prices, and I know that's going to be true 10 years from now. They want fast delivery; they want vast selection. It's impossible to imagine a future 10 years from now where a customer comes up and says, 'Jeff I love Amazon; I just wish the prices were a little higher,' or 'I love Amazon; I just wish you'd deliver a little more slowly.' Impossible."

—Medium, April 5, 2018

45. "Amazon.com strives to be the e-commerce destination where consumers can find and discover anything they want to buy online."

—Medium, April 5, 2018

46. "When you collect and store data, you need to be clear. And not just in subsection 17, paragraph 3. You need to be obviously clear. It's why Amazon greets users by name, so they know they are not anonymous."

—Recode's Code conference, May 31, 2016

47. "We don't do PowerPoint (or any other slide-oriented) presentations at Amazon. Instead, we write narratively structured six-page memos. We silently read one at the beginning of each meeting in a kind of 'study hall'."

—Jeff Bezos's 2017 Letter to Amazon Shareholders, April 18, 2018

48. "We are culturally pioneers. We like to disrupt even our own business. Other companies have different cultures and sometimes don't like to do that. Our job is to bring those industries along."

—Wired, November 13, 2011

49. "Amazon today remains a small player in global retail. We represent a low single-digit percentage of the retail market, and there are much larger retailers in every country where we operate. And that's largely because nearly 90% of retail remains offline, in brick and mortar stores."

—Jeff Bezos's 2018 Letter to Amazon Shareholders, April 11, 2019

50. "Our success at Amazon is a function of how many experiments we do per year, per month, per week, per day."

—Observer, November, 20, 2017

51. “We love to invent. We love to pioneer. We even like going down alleys that turn out to be blind alleys. Every once in a while one of those blind alleys opens up into a broad avenue, and that's really fun.”

—CNBC, September 6, 2012

52. “We’ve had three big ideas at Amazon that we’ve stuck with for 18 years, and they’re the reason we’re successful: Put the customer first. Invent. And be patient.”

—The Washington Post, September 3, 2013

53. “I like the rate of change, I like the opportunity, the “canvas” upon which we can invent new things. And I am not alone in Amazon in that regard. We have a big team of people who are genetically predisposed to have fun while inventing new things. It’s sort of in our DNA.”

—Fortune, June 29, 2010

54. “Companies have short life spans. And Amazon will be disrupted one day.... I don't worry about it because I know it's inevitable. Companies come and go. And the companies that are the shiniest and most important of any era, you wait a few decades and they're gone.... I would love for it to be after I'm dead.”

—CBS News, December 1, 2013

AWS

55. “On the other side of Amazon's online retailing is a business customers know little about. It's called Amazon Web Services, AWS, and may soon become Amazon's biggest business.

To keep track of its massive online orders, Amazon built a large and sophisticated computing infrastructure. Amazon figured out it could also expand that infrastructure to store data and run websites for hundreds of thousands of outside companies and government agencies on what is known as the cloud.”

—CBS News, December 1, 2013

56. “The biggest needle movers will be things that customers don’t know to ask for. We must invent on their behalf. We have to tap into our own inner imagination about what’s possible.

AWS itself – as a whole – is an example. No one asked for AWS. No one. Turns out the world was in fact ready and hungry for an offering like AWS but didn’t know it.”

—Jeff Bezos’s 2018 Letter to Amazon Shareholders, April 11, 2019

57. “Most Internet startups and a lot of big Internet companies run on top of AWS. Netflix very famously, and you could say, ‘Oh that's very odd because Netflix in a way is a competitor of Amazon’.”

—CBS News, December 1, 2013

Amazon Studios

58. “When we win a Golden Globe, it helps us sell more shoes.”

—The Atlantic, November, 2019

59. “We don’t compete with Netflix. I think people are going to subscribe to both.”

—Recode’s Code conference, May 31, 2016

Blue Origin

60. “We are a small team committed to seeding an enduring human presence in space. Blue will pursue this long-term objective patiently.”

—Forbes, September 3, 2019

61. “Our motto is, ‘Gradatim Ferociter’. It means step by step ferociously.... Basically, you can’t skip steps, you have to put one foot in front of the other, things take time, there are no shortcuts, but you want to do those steps with passion and ferocity.”

—Inc.com, May 30, 2019

62. “I believe and I get increasing conviction with every passing year, that Blue Origin, the space company, is the most important work that I’m doing. And so there is a whole plan for Blue Origin.”

—Business Insider, April 28, 2018

63. “Right now, the business model with Blue Origin is I sell Amazon stock. I’m willing to be patient for decades.”

—Wired, October 15, 2018

64. "Nobody gets into the space business because they've done an exhaustive analysis of all the industries they might invest in and they find that the one with the least risk and the highest returns on capital is the space business."

—Fortune, March 24, 2016

65. "Blue Origin is expensive enough to be able to use that fortune. I am liquidating about $1 billion a year of Amazon stock to fund Blue Origin. And I plan to continue to do that for a long time....

I am very lucky that I feel like I have a mission-driven purpose with Blue Origin that is incredibly important for civilization long term. And I am going to use my financial lottery winnings from Amazon to fund that."

—Business Insider, April 28, 2018

66. "I want to see millions of people living and working in space. I think it's important. I also just love it. I love change. I love technology. I love the engineers we have. They're brilliant."

—Business Insider, December 13, 2014

Space Colonization

67. "The Earth is finite, and if the world economy and population is to keep expanding, space is the only way to go."

—The Atlantic, May 10, 2019

68. "The reason we go to space is to save the Earth.... We need to move industry to the moon. Earth will be zoned for residential and light industry."

—re:MARS conference in Las Vegas, June 6, 2019

69. "We can have a trillion humans in the solar system, which means we'd have a thousand Mozarts and a thousand Einsteins. This would be an incredible civilization."

—The Atlantic, November, 2019

70. "Eventually it will be much cheaper and simpler to make really complicated things like microprocessors ... in space and then send those highly complex manufactured objects back down to Earth so that we don't have the big factories and pollution-generating industries that make those things now."

—CBS Evening News, July 16, 2019

71. "The moon is an essential place to get resources to build these kind of O'Neill-style colonies. They're not what you imagine. I mean, they'll have farms and rivers and universities; they could have a million people in them. They're cities. But I'd want to be able to come back and forth to Earth too."

—Wired, October 15, 2018

72. "Ultimately, we're going to be able to get hydrogen from that water on the moon, and be able to refuel these vehicles on the surface of the moon."

—The Atlantic, May 10, 2019

73. "If I wanted to buy tourist trips to fly to the International Space Station and Soyuz and those things, there's nothing wrong with that. But that's $35 million. I want to lower the cost of access to space."

—Wired, November 13, 2011

74. "The key is reusability. This civilization I'm talking about of getting comfortable living and working in space and having millions of people and then billions of people and then finally a trillion people in space. You can't do that with space vehicles that you use once and then throw away. It's a ridiculous, costly way to get into space."

—Insider, April 30, 2018

75. “People are going to want to live on Earth and they’re going to want to live off Earth – there are going to be very nice places to live off Earth as well.... People will make that choice.”

—CBS Evening News, July 16, 2019

76. “I’m interested in space, because I’m passionate about it. I’ve been studying it and thinking about it since I was a 5-year-old boy. But that is not why I’m pursuing this work. I’m pursuing this work, because I believe if we don’t we will eventually end up with a civilization of stasis, which I find very demoralizing.

I don’t want my great-grandchildren’s great-grandchildren to live in a civilization of stasis. We all enjoy a dynamic civilization of growth and change.”

—Business Insider, April 28, 2018

77. “One day I’m going to be the first one to have an amusement park on the moon.”

—Wired, October 15, 2018

78. "The kids here ["Club for the Future" powered by Blue Origin] and their kids and grandchildren will build these colonies. My generation's job is to build the infrastructure so they will be able to. We are going to build a road to space....

And then amazing things will happen. Then you'll see entrepreneurial creativity. Then you'll see space entrepreneurs start companies in their dorm rooms. That can't happen today."

—Fortune, May 9, 2019

79. "[I studied alternative means of propulsion] for three years with a small group of brilliant people and concluded with complete conviction that rockets are actually not just a good solution for getting off the surface of the earth, they're a great solution. But they have to be reusable."

—Wired, October 15, 2018

80. "We will settle Mars. And we should, because it's cool."

—Fortune, June 1, 2016

81. "Getting to Mars is only one of my goals. For Elon [Musk], it's the only one."

—Recode's Code conference, May 31, 2016

82. “You can’t predict how or why or when — but new worlds have a way of saving old worlds. That’s how it should be. We need the frontier. We need the people moving out into space.”

—Business Insider, December 13, 2014

The Washington Post

83. "I would not have bought the Washington Post if it had been a financially upside-down salty-snack-food company."

—Fortune, March 24, 2016

84. "This is the first company I've ever been involved with on a large scale that I didn't build from scratch.... I did no due diligence, and I did not negotiate with Don [Graham, who has been publicly mum on the Post since selling it]. I just accepted the number he proposed."

—Fortune, March 24, 2016

85. "There is one thing I'm certain about: there won't be printed newspapers in twenty years."

—The New Yorker, August 5, 2013

86. "It's important for The Post not just to survive, but to grow. The product of The Post is still great. The piece that's missing is that it's a challenged business. No business can continue to shrink. That can only go on for so long before irrelevancy sets in."

—The Washington Post, September 3, 2013

Part 3: SUCCESS LESSONS

Being Jeff Bezos

87. "I'm a genetic optimist. I've been told, 'Jeff, you're fooling yourself; the problem is unsolvable.' But I don't think so. It just takes a lot of time, patience and experimentation."

—The Washington Post, September 3, 2013

88. "I'm a very goofy sort of person in many ways. I tell duff jokes. I kind of have to grow on somebody. I'm not the kind of person you meet and kind of go, 'Oh I love this guy!' and it's great and you go home and tell all your girlfriends about him. That's not me."

—The Guardian, February 11, 2001

89. "I don't associate with any party.... I do have things I care about and some of those are public, like gay marriage. But I don't feel the need to have an opinion on every issue."

—The Washington Post, September 4, 2013

90. "I am a kind of American exceptionalist.... I believe this is a good country. But our elected officials are not perfect. Our regulators are not perfect."

—The Washington Post, September 4, 2013

91. "One of my jobs is to encourage people to be bold. It's incredibly hard. Experiments are, by their very nature, prone to failure. A few big successes compensate for dozens and dozens of things that didn't work.

Bold bets — Amazon Web Services, Kindle, Amazon Prime, our third-party seller business — all of those things are examples of bold bets that did work, and they pay for a lot of experiments."

—Business Insider, December 13, 2014

Strategy Development

92. “Base your strategy on things that won’t change.”

—Forbes, April 4, 2012

93. “Maintain a firm grasp of the obvious at all times.”

—Wired, November 13, 2011

Business Plans & Reality

94. "Things never go smoothly."

—The Academy of Achievement, May 4, 2001

95. "It's very important for entrepreneurs to be realistic. So if you believe on that first day while you're writing the business plan that there's a 70 percent chance that the whole thing will fail, then that kind of relieves the pressure of self-doubt."

—Medium, July 4, 2013

96. "Any business plan won't survive its first encounter with reality. The reality will always be different. It will never be the plan."

—Forbes, September 23, 2013

Motivation

97. "The thing that motivates me is a very common form of motivation. And that is, with other folks counting on me, it's so easy to be motivated."

—theDECALIFE, January 12, 2019

98. "I strongly believe that missionaries make better products. They care more. For a missionary, it's not just about the business. There has to be a business, and the business has to make sense, but that's not why you do it. You do it because you have something meaningful that motivates you."

—Fortune, June 29, 2010

99. "Focusing on customer needs is very motivating because customers will never be satisfied."

—The Hollywood Reporter, July 15, 2015

Change Your Mind

100. "The smartest people are constantly revising their understanding, reconsidering a problem they thought they'd already solved. They're open to new points of view, new information, new ideas, contradictions, and challenges to their own way of thinking."

—Inc., February 15, 2019

101. "People who were right a lot of the time were people who often changed their minds.... It's perfectly healthy — encouraged, even — to have an idea tomorrow that contradicted your idea today."

—CNBC, March 12, 2019

Innovation

102. “Frugality drives innovation, just like other constraints do. One of the only ways to get out of a tight box is to invent your way out.”

—Bloomberg, April 17, 2008

103. “You have to be willing to be misunderstood if you’re going to innovate.”

—Inc., May 7, 2014

104. “You’d be pretty depressed right now because the last nugget of gold would be gone. But the good thing is, with innovation, there isn’t a last nugget. Every new thing creates two new questions and two new opportunities.”

—TED Talks, 2003

Decision Making

105. "Where you are going to spend your time and your energy is one of the most important decisions you get to make in life."

—Business Insider, December 13, 2014

106. "There are many decisions that we can make with math. For those kinds of problems, if you used gut intuition, that would be foolish. But a lot of the decisions that you have to make around consumers are not that kind of thing. You can't put into a spreadsheet how people are going to behave around a new product."

—US News, November 19, 2008

107. "These decisions must be made methodically, carefully, slowly and with great deliberation and consultation. If you walk through and don't like what you see on the other side, you can't get back to where you were before."

—The New York Times, March 2, 2019

108. "For every leader in the company, not just for me, there are decisions that can be made by analysis. These are the best kinds of decisions! They're fact-based decisions."

—Fast Company, August 1, 2004

109. "The great thing about fact-based decisions is that they overrule the hierarchy. The most junior person in the company can win an argument with the most senior person with regard to a fact-based decision. For intuitive decisions, on the other hand, you have to rely on experienced executives who've honed their instincts."

—Inc., May 7, 2014

110. "Every well-intentioned, high-judgment person we asked told us not to do it. We got some good advice, we ignored it, and it was a mistake. But that mistake turned out to be one of the best things that happened to the company."

—Forbes, Apr 30, 2013

111. "The framework I found which made the decision incredibly easy was what I called – which only a nerd would call – a 'regret minimization framework'.

So I wanted to project myself forward to age 80 and say, 'Okay, now I'm looking back on my life. I want to have minimized the number of regrets I have.' (If I did not start Amazon) I knew that that would haunt me every day."

—Fast Company, August 1, 2004

Long Term-Thinking

112. "My own view is that every company requires a long-term view."

—Medium, February 21, 2018

113. "If you're going to take a long-term orientation, you have to be willing to stay heads down and ignore a wide array of critics, even well-meaning critics. If you don't have a willingness to be misunderstood for a long period of time, then you can't have a long-term orientation."

—US News, November 19, 2008

114. "The three big ideas at Amazon are long-term thinking, customer obsession, and a willingness to invent."

—Fortune, November 16, 2012

115. “You can do the math 15 different ways, and every time the math tells you that you shouldn’t lower prices because you’re gonna make less money. That’s undoubtedly true in the current quarter, in the current year.

But it’s probably not true over a 10-year period, when the benefit is going to increase the frequency with which your customers shop with you, the fraction of their purchases they do with you as opposed to other places. Their overall satisfaction is going to go up.”

—Fast Company, August 1, 2004

116. “What we’re really focused on is thinking long-term, putting the customer at the center of our universe and inventing. Those are the three big ideas to think long-term because a lot of invention doesn’t work.”

—Product Hunt, May 5, 2019

117. “Sometimes we measure things and see that in the short term they actually hurt sales. But we do it anyway, because we believe that the short-term metrics probably aren’t indicative of the long term.”

—Forbes, May 4, 2013

118. “[If humans] think long term, we can accomplish things that we wouldn’t otherwise accomplish.”

—The Atlantic, November, 2019

Failure & Risk

119. "If you want to be inventive, you have to be willing to fail."

—Forbes, April 4, 2012

120. "The good news for shareowners is that a single big winning bet can more than cover the cost of many losers."

—Jeff Bezos's 2018 Letter to Amazon Shareholders, April 11, 2019

121. "If you have a business idea with no risk, it's probably already being done. You've got to have something that might not work. It will be, in many ways, an experiment. We take risks all the time, we talk about failure."

—re:MARS conference in Las Vegas, June 6, 2019

122. “We need big failures in order to move the needle. If we don’t, we’re not swinging enough. You really should be swinging hard, and you will fail, but that’s okay. When you swing hard in baseball, the maximum number of runs you can get is four. When you swing hard in business, you can get 100 runs.

So you really should swing hard, and you are going to fail a lot, and you need a culture that supports that. When do you know when to stop? When the last champion is ready to throw in the towel, it’s time to stop. And usually, that last champion is me.”

—Fast Company, June 6, 2019

123. “You’ve gotta accept that your business is going to be in many ways an experiment, and it might fail. That’s OK. That’s what risk is.”

—Fortune, June 7, 2019

124. “As a company grows, everything needs to scale, including the size of your failed experiments. If the size of your failures isn’t growing, you’re not going to be inventing at a size that can actually move the needle.”

—Jeff Bezos’s 2018 Letter to Amazon Shareholders, April 11, 2019

125. “You absolutely have to be willing to make something that’s going to be remarkable. You’ve got to take some risks.”

—The Hollywood Reporter, July 15, 2015

Experiments & Invention

126. "Failure and invention are inseparable twins."

—Entrepreneur Magazine, May 24, 2017

127. "It's not an experiment if you know it's going to work."

—theDECALIFE, January 12, 2019

128. "In some cases, things are inevitable. The hard part is that you don't know how long it might take, but you know it will happen if you're patient enough. Ebooks had to happen. Infrastructure web services had to happen."

—Wired, November 13, 2011

129. "As a company, one of our greatest cultural strengths is accepting the fact that if you're going to invent, you're going to disrupt. A lot of entrenched interests are not going to like it."

—Wired, November 13, 2011

130. “The thing about inventing is you have to be both stubborn and flexible, more or less simultaneously. If you’re not stubborn, you’ll give up on experiments too soon. And if you’re not flexible, you’ll pound your head against the wall and you won’t see a different solution to a problem you’re trying to solve.”

—Medium, July 4, 2013

131. “If you double the number of experiments you do per year you’re going to double your inventiveness.”

—Inc., May 7, 2014

132. “When we’re at our best, we don’t wait for external pressures. We are internally driven to improve our services, adding benefits and features, before we have to. We lower prices and increase value for customers before we have to. We invent before we have to.”

—Jeff Bezos’s 2012 Letter to Amazon Shareholders, April, 2013

133. “There’ll always be serendipity involved in discovery.”

—Medium, July 4, 2013

134. “It’s probably a survival skill that we’re curious and like to explore. Our ancestors, who were incurious and failed to explore, probably didn’t live as long as the ones who were looking over the next mountain range to see if there were more sources of food and better climates and so on and so on.”

—Business Insider, December 13, 2014

135. “We like to pioneer, we like to explore, we like to go down dark alleys and see what's on the other side.”

—CBS News, December 1, 2013

136. “We are willing to go down a bunch of dark passageways, and occasionally we find something that really works.”

—Medium, July 4, 2013

Pioneering Requires Being Misunderstood

137. "If you never want to be criticized, for goodness' sake don't do anything new."

—CNBC, April 29, 2014

138. "We are willing to be misunderstood for long periods of time."

—Forbes, April 4, 2012

139. "Beautiful speech doesn't need protection. Ugly speech needs protection."

—Bloomberg, June 1, 2016

140. "Any time you do something big, that's disruptive — Kindle, Amazon Web Services — there will be critics. And there will be at least two kinds of critics. There will be well-meaning critics who genuinely misunderstand what you are doing or genuinely have a different opinion.

And there will be the self-interested critics that have a vested interest in not liking what you are doing and they will have reason to misunderstand. And you have to be willing to ignore both types of critics. You listen to them, because you want to see, always testing, is it possible they are right?"

—GeekWire, June 7, 2011

141. “The best defense to speech that you don’t like about yourself as a public figure is to develop a thick skin. It’s really the only effective defense.”

—Recode’s Code conference, May 31, 2016

142. “If you can’t tolerate critics, don’t do anything new or interesting.”

—theDECALIFE, January 12, 2019

143. “My approach to criticism and what I teach and preach inside Amazon — is when you’re criticized, first look in a mirror and decide, are your critics right? If they’re right, change. Don’t resist.”

—Business Insider, April 28, 2018

Part 4: BRILLIANT ADVICE FOR ENTREPRENEURS

Putting the Customer First

144. "Obsess over customers."

—Forbes, April 4, 2012

145. "Our version of a perfect customer experience is one in which our customer doesn't want to talk to us. Every time a customer contacts us, we see it as a defect. I've been saying for many, many years, people should talk to their friends, not their merchants.

And so we use all of our customer service information to find the root because of any customer contact. What went wrong? Why did that person have to call? Why aren't they spending that time talking to their family instead of talking to us? How do we fix it?"

—Wired, November 13, 2011

146. "If you're competitor-focused, you have to wait until there is a competitor doing something. Being customer-focused allows you to be more pioneering."

—US News, November 19, 2008

147. “If there’s one reason we have done better than all of our peers in the Internet space over the last six years, it is because we have focused like a laser on customer experience, and that really does matter in any business.”

—Medium, July 4, 2013

148. “I was in a hotel where they had three little bottles on the counter. One was named ‘Shine’, and one was named ‘Smooth’, and one was named something else. One of them was hand lotion, one was soap, and one was shampoo. It was a puzzle-solving exercise to figure out which was which.

They are indulging the designer’s own desire to be creative, rather than trying to figure out, ‘Okay, how can I make this as easy for the user as possible? I don’t want to use my creative energy on somebody else’s user interface’.”

—Esquire, September 25, 2008

149. “There are two ways to extend a business. Take inventory of what you are good at and extend out from your skills. Or determine what your customers need and work backwards, even if it requires learning new skills.”

—Medium, July 4, 2013

150. "When we first did customer reviews 20 years ago, some book publishers were not happy about it because some of [the reviews] were negative. So it was a very controversial practice at that time, but we thought it was right, and so we stuck to our guns and had a deep keel on that and didn't change."

—CNBC, May 17, 2018

151. "The best customer service is if the customer doesn't need to call you, doesn't need to talk to you. It just works."

—Medium, July 4, 2013

152. "The most important single thing is to focus obsessively on the customer. Our goal is to be earth's most customer-centric company."

—Medium, February 21, 2018

153. "Customers are very smart. You should never underestimate customers."

—Business Insider, April 28, 2018

154. "We see our customers as invited guests to a party, and we are the hosts. It's our job every day to make every important aspect of the customer experience a little bit better."

—Forbes, Apr 30, 2013

155. “You have to use your judgment. In cases like that, we say, ‘Let’s be simpleminded. We know this is a feature that’s good for customers. Let’s do it’.”

—Fast Company, August 1, 2004

156. “In the long run, if you take care of customers, that is taking care of shareholders. We do price elasticity studies. And every time the math tells us to raise prices.... [D]oing so would erode trust. And that erosion of trust would cost us much more in the long term.”

—CBS News, December 1, 2013

Choosing Slim Profits

157. “Your margin is my opportunity.”

—Fortune, November 16, 2012

158. “Tech companies always have high margins, except for Amazon. We’re the only tech company with low margins.... We really obsess over small defects. That’s what drives up costs. Because the most expensive thing you can do is make a mistake.”

—Wired, November 13, 2011

159. “There are two kinds of companies, those that work to try to charge more and those that work to charge less. We will be the second.”

—BBC, August 6, 2013

160. “We work very, very hard to be able to afford to offer customers low margins. We’d rather have a very large customer base and low margins than a smaller customer base and higher margins.”

—Wired, November 13, 2011

How to Build a Successful Company

161. "There are always shiny things. A company shouldn't get addicted to being shiny, because shiny doesn't last. You really want something that's much deeper-keeled. You want your customers to value your service. And there are companies that haven't gone through tough times, so they're not really tested."

—Entrepreneur Magazine, May 24, 2017

162. "All businesses need to be young forever. If your customer base ages with you, you're Woolworth's. The number one rule has to be: Don't be boring."

—The Washington Post, September 4, 2013

163. "There are two ways to build a successful company. One is to work very, very hard to convince customers to pay high margins. The other is to work very, very hard to be able to afford to offer customers low margins. They both work.

We're firmly in the second camp. It's difficult — you have to eliminate defects and be very efficient. But it's also a point of view. We'd rather have a very large customer base and low margins than a smaller customer base and higher margins."

—Wired, November 13, 2011

164. “If you only do things where you know the answer in advance, your company goes away.”

—Esquire, September 25, 2008

165. “I very frequently get the question: ‘What’s going to change in the next 10 years?’ And that is a very interesting question; it’s a very common one. I almost never get the question: ‘What's not going to change in the next 10 years?’

And I submit to you that that second question is actually the more important of the two – because you can build a business strategy around the things that are stable in time.... [I]n our retail business, we know that customers want low prices, and I know that's going to be true 10 years from now. They want fast delivery; they want vast selection.”

—Inc., November 6, 2017

166. “[I]nside our culture, we understand that even though we have some big businesses, new businesses start out small.... The biggest oak starts from an acorn and if you want to do anything new, you’ve got to be willing to let that acorn grow into a little sapling and then finally into a small tree and maybe one day it will be a big business on its own.”

—Product Hunt, May 5, 2019

167. “If you don’t understand the details of your business you are going to fail.”

—theDECALIFE, January 12, 2019

Marketing & Advertising

168. "In the old world, you devoted 30% of your time to building a great service and 70% of your time to shouting about it. In the new world, that inverts."

—Adweek, September 14, 2009

169. "The balance of power is shifting toward consumers and away from companies... The right way to respond to this if you are a company is to put the vast majority of your energy, attention and dollars into building a great product or service and put a smaller amount into shouting about it, marketing it."

—Inc., May 7, 2014

Brand-Building

170. “Your brand is what people say about you when you’re not in the room.”

—Inc., February 15, 2019

171. “A brand for a company is like a reputation for a person. You earn reputation by trying to do hard things well.”

—Bloomberg, August 2, 2004

Hiring Lessons

172. "What characteristics do I look for when hiring somebody? That's one of the questions I ask when interviewing. I want to know what kind of people they would hire."

—Esquire, January, 2002

173. "I'd rather interview 50 people and not hire anyone than hire the wrong person."

—Fast Company, August 1, 2004

174. "Everyone has to be able to work in a call center."

—Forbes, April 4, 2012

175. "We approach our recruiting in unapologetically elitist fashion."

—The Atlantic, November, 2019

176. "You can work long, hard or smart, but at Amazon.com you can't choose two out of three.... It's not easy to work here."

—The New York Times, August 15, 2015

177. "We pay very low cash compensation relative to most companies. We also have no incentive compensation of any kind. And the reason we don't is because it is detrimental to teamwork."

—Fortune, November 16, 2012

178. "You're going to have a willingness to repeatedly fail if you're going to experiment. For a certain kind of person, that is a very exciting, very motivating culture. So, we attract those kinds of people."

—US News, November 19, 2008

Books, Gadgets & Social Networking

179. "People don't want gadgets, they want services."

—CNBC, September 6, 2012

180. "Attention is the scarce commodity of the late 20th century."

—Esquire, September 25, 2008

181. "We want to make money when people use our devices, not when people buy our devices."

—CNBC, September 6, 2012

182. "Books don't just compete against books. Books compete against people reading blogs and news articles and playing video games and watching TV and going to see movies.

Books are the competitive set for leisure time. It takes many hours to read a book. It's a big commitment. If you narrow your field of view and only think about books competing against books, you make really bad decisions. What we really have to do, if we want a healthy culture of long-form reading, is to make books more accessible."

—Business Insider, December 13, 2014

183. "For better or worse, it is really not a part of our culture to look at things defensively. We rarely say, 'Oh my God, we've got to do something about that existential threat.' Maybe one day we'll become extinct because of that deficiency in our nature. I don't know.

We look at things through a different lens. We say, 'Oh, here's this incredible phenomenon called social networking. How can we be inspired by that to make our business better?' I hope we find something."

—Wired, November 13, 2011

184. "We want people to feel like they're visiting a place, rather than a software application."

—Wired, March 1, 1999

Internet

185. "The Internet is disrupting every media industry, people can complain about that, but complaining is not a strategy."

—CBS News, December 1, 2013

186. "This is Day 1 for the Internet. We still have so much to learn."

—Forbes, April 4, 2012

187. "The Internet took a lot away from publishing, but one of the gifts it brought was that distribution is basically free. No more printing presses needed."

—Recode's Code conference, May 31, 2016

188. "You can see going forward that Internet, access to broadband is going to be very close to being a fundamental human need as we move forward."

—re:MARS conference in Las Vegas, June 6, 2019

189. "I think of the Internet like this big, new, powerful technology. It's horizontal. It affects every industry. And if you think of it even more broadly, it's tech and machine learning, big data, and all these kinds of things. These are big, horizontal, powerful technologies....

But we're also finding out that these powerful tools enable some very bad things, too, like letting authoritarian governments interfere in free democratic elections in the world. This is an incredibly scary thing."

—Business Insider, April 28, 2018

190. "We are at the 1908 Hurley washing machine stage with the Internet. That's where we are. We don't get our hair caught in it, but that's the level of primitiveness of where we are. We're in 1908.... And I do think there's more innovation ahead of us than there is behind us."

—TED Talks, 2003

191. "Sometimes I think the Time Person of the Year is chosen for the man, and I think sometimes they are chosen as a symbol of something, and my selection was clearly that. They weren't choosing Jeff Bezos so much as they were choosing me as a symbol for the Internet. Yeah, my parents were very proud. I mean, they are parents. They are not objective."

—Esquire, January, 2002

Productivity Tips for Entrepreneurs

192. "Go to bed early and wake up early. The morning hours are good."

—Esquire, January, 2002

193. "Life's too short to hang out with people who aren't resourceful."

—Entrepreneur Magazine, May 24, 2017

Big Winners

194. "You're still going to be wrong nine times out of ten. In business, every once in a while, when you step up to the plate, you can score 1,000 runs. This long-tailed distribution of returns is why it's important to be bold. Big winners pay for the many previous experiments."

—Forbes, May 7, 2019

195. "You should give up on things, but you shouldn't give up easily."

—Recode's Code conference, May 31, 2016

196. "When it's tough, will you give up, or will you be relentless?"

—Entrepreneur Magazine, May 24, 2017

197. "If you decide that you're going to do only the things you know are going to work, you're going to leave a lot of opportunity on the table."

—Inc., May 7, 2014

198. "We can't go backwards. We also can't think small. We need to think big and lean into the future. The death knell for any enterprise is to glorify the past no matter how good it was."

—The Washington Post, September 4, 2013

199. “Part of company culture is path-dependent – it’s the lessons you learn along the way. One piece of the culture here that is true of my personality is that I have never believed that you couldn’t be serious and have fun at the same time. It’s perhaps most important to have fun when stumbling. It is harder.”

—US News, November 19, 2008

FREE book

For a limited time Olivia Longray is giving away the book called “A Simple Weight Loss Plan That Can Work for You: How to Lose Weight Quickly in an Atmosphere of Love.”

This is your only chance to get it FREE (no strings attached).

Here is the link for you: http://bit.ly/2xkTiKY

Enjoy!

Your Reviews

If the book proves useful and appealing to you, please, leave a short review on the book's Amazon page and give me your thoughts on it! Your opinion will surely help other readers to decide if my book is worth their time and money. Tell others of your impressions and my thanks to you for it in advance!

Have a great day!

Olivia Longray

A Note on Sources

1. Henry Blodget, "I Asked Jeff Bezos The Tough Questions — No Profits, The Book Controversies, The Phone Flop — And He Showed Why Amazon Is Such A Huge Success," Business Insider, December 13, 2014, https://www.businessinsider.com/amazons-jeff-bezos-on-profits-failure-succession-big-bets-2014-12#ixzz3gdJa8T9C
2. "Top 10 Jeff Bezos Quotes," theDECALIFE, January 12, 2019, https://thedecalife.com/top-10-jeff-bezos-quotes/
3. Cal Fussman, "Jeff Bezos: What I've Learned," Esquire, January, 2002, https://www.esquire.com/news-politics/interviews/a2033/esq0102-jan-bezos/
4. Mathias Döpfner, "Jeff Bezos reveals what it's like to build an empire and become the richest man in the world — and why he's willing to spend $1 billion a year to fund the most important mission of his life," Business Insider, April 28, 2018, https://www.businessinsider.com/jeff-bezos-interview-axel-springer-ceo-amazon-trump-blue-origin-family-regulation-washington-post-2018-4
5. Cal Fussman, "Jeff Bezos: What I've Learned," Esquire, January, 2002, https://www.esquire.com/news-politics/interviews/a2033/esq0102-jan-bezos/
6. Steven Levy, "Jeff Bezos Owns the Web in More Ways Than You Think," Wired, November 13, 2011, https://www.wired.com/2011/11/ff_bezos/2
7. Franklin Foer, "Jeff Bezos's Master Plan," The Atlantic, November, 2019, https://www.theatlantic.com/magazine/archive/2019/11/what-jeff-bezos-wants/598363/
8. Rose Leadem, "10 Inspiring Quotes From the Fearless Jeff Bezos," Entrepreneur Magazine, May 24, 2017, https://www.entrepreneur.com/slideshow/294490
9. Andrew Smith, "Brought to book," The Guardian, February 11, 2001,

https://www.theguardian.com/books/2001/feb/11/computingandthenet.technology
10. Adam Lashinsky, "Bezos Prime," Fortune, March 24, 2016, https://fortune.com/longform/amazon-jeff-bezos-prime/
11. Mathew J Maniyamkott, "Inspirational quotes from Jeff Bezos on how to court success in the long term," YourStory, January, 12, 2018, https://yourstory.com/2018/01/jeff-bezos-quotes-for-success
12. "Top 10 Jeff Bezos Quotes," theDECALIFE, January 12, 2019, https://thedecalife.com/top-10-jeff-bezos-quotes/
13. Jeff Bezos, "2017 Letter to Shareholders," April 18, 2018, https://blog.aboutamazon.com/company-news/2017-letter-to-shareholders/
14. Ivan Minic, "10 Business Lessons From Jeff Bezos," Medium, July 4, 2013, https://medium.com/@burek/10-business-lessons-from-jeff-bezos-3fe167f58d65
15. Steven Levy, "Jeff Bezos Wants Us All to Leave Earth—for Good," Wired, October 15, 2018, https://www.wired.com/story/jeff-bezos-blue-origin/
16. "Jeff Bezos fireside chat at re:MARS 2019", Amazon News, June 6, 2019, https://www.youtube.com/watch?v=AbpXSM8WW4s
17. Cal Fussman, "Jeff Bezos: What I've Learned," Esquire, January, 2002, https://www.esquire.com/news-politics/interviews/a2033/esq0102-jan-bezos/
18. Alan Deutschman, "Inside the Mind of Jeff Bezos," Fast Company, August 1, 2004, https://www.fastcompany.com/90435429/this-50-device-is-trying-to-finally-kill-off-the-walkie-talkie
19. Henry Blodget, "I Asked Jeff Bezos The Tough Questions — No Profits, The Book Controversies, The Phone Flop — And He Showed Why Amazon Is Such A Huge Success," Business Insider, December 13, 2014, https://www.businessinsider.com/amazons-jeff-bezos-on-profits-failure-succession-big-bets-2014-12#ixzz3gdJa8T9C

20. Cal Fussman, "Jeff Bezos: What I've Learned," Esquire, January, 2002, https://www.esquire.com/news-politics/interviews/a2033/esq0102-jan-bezos/
21. Jillian D'Onfro, "17 quotes that show how Jeff Bezos turned Amazon into a $200 billion company over 20 years," Business Insider, July 16, 2015, https://www.businessinsider.com/amazon-ceo-jeff-bezos-quotes-2015-7
22. Adam Lashinsky, "Bezos Prime," Fortune, March 24, 2016, https://fortune.com/longform/amazon-jeff-bezos-prime/
23. Mathias Döpfner, "Jeff Bezos reveals what it's like to build an empire and become the richest man in the world — and why he's willing to spend $1 billion a year to fund the most important mission of his life," Business Insider, April 28, 2018, https://www.businessinsider.com/jeff-bezos-interview-axel-springer-ceo-amazon-trump-blue-origin-family-regulation-washington-post-2018-4
24. Henry Blodget, "I Asked Jeff Bezos The Tough Questions — No Profits, The Book Controversies, The Phone Flop — And He Showed Why Amazon Is Such A Huge Success," Business Insider, December 13, 2014, https://www.businessinsider.com/amazons-jeff-bezos-on-profits-failure-succession-big-bets-2014-12#ixzz3gdJa8T9C
25. Mathias Döpfner, "Jeff Bezos reveals what it's like to build an empire and become the richest man in the world — and why he's willing to spend $1 billion a year to fund the most important mission of his life," Business Insider, April 28, 2018, https://www.businessinsider.com/jeff-bezos-interview-axel-springer-ceo-amazon-trump-blue-origin-family-regulation-washington-post-2018-4
26. Jason Del Rey, "Five Things We Learned About Amazon CEO Jeff Bezos From His Rare Public Appearance," Vox, December 3, 2014, https://www.vox.com/2014/12/3/11633474/five-things-we-learned-about-amazon-ceo-jeff-bezos-from-his-rare
27. Cal Fussman, "Jeff Bezos: What I've Learned," Esquire, January, 2002,

https://www.esquire.com/news-politics/interviews/a2033/esq0102-jan-bezos
28. Henry Blodget, "I Asked Jeff Bezos The Tough Questions — No Profits, The Book Controversies, The Phone Flop — And He Showed Why Amazon Is Such A Huge Success," Business Insider, December 13, 2014, https://www.businessinsider.com/amazons-jeff-bezos-on-profits-failure-succession-big-bets-2014-12#ixzz3gdJa8T9C
29. Cal Fussman, "Jeff Bezos: What I've Learned," Esquire, January, 2002, https://www.esquire.com/news-politics/interviews/a2033/esq0102-jan-bezos/
30. Cal Fussman, "Jeff Bezos: What I've Learned," Esquire, January, 2002, https://www.esquire.com/news-politics/interviews/a2033/esq0102-jan-bezos/
31. Cal Fussman, "Jeff Bezos: What I've Learned," Esquire, January, 2002, https://www.esquire.com/news-politics/interviews/a2033/esq0102-jan-bezos/
32. Cal Fussman, "Jeff Bezos: What I've Learned," Esquire, January, 2002, https://www.esquire.com/news-politics/interviews/a2033/esq0102-jan-bezos/
33. Cal Fussman, "Jeff Bezos: What I've Learned," Esquire, January, 2002, https://www.esquire.com/news-politics/interviews/a2033/esq0102-jan-bezos/
34. Cal Fussman, "Jeff Bezos: What I've Learned," Esquire, January, 2002, https://www.esquire.com/news-politics/interviews/a2033/esq0102-jan-bezos/
35. "Amazon founder Jeff Bezos delivers Princeton University's 2010 Baccalaureate address," Princeton University, 2010, https://www.youtube.com/watch?v=Duml1SHJqNE
36. "The Inner Bezos," March 01, 1999, https://www.wired.com/1999/03/bezos-3/
37. Charlie Rose, "Amazon's Jeff Bezos looks to the future," CBS News, December 1, 2013, https://www.cbsnews.com/news/amazons-jeff-bezos-looks-to-the-future/

38. "King of Cyber-Commerce," The Academy of Achievement, May 4, 2001, https://www.achievement.org/achiever/jeffrey-p-bezos/
39. "Top 10 Jeff Bezos Quotes," theDECALIFE, January 12, 2019, https://thedecalife.com/top-10-jeff-bezos-quotes/
40. Cal Fussman, "Jeff Bezos: What I've Learned," Esquire, January, 2002, https://www.esquire.com/news-politics/interviews/a2033/esq0102-jan-bezos/
41. "Jeff Bezos nails it on how to succeed in business," Product Hunt, May 5, 2019, https://twitter.com/producthunt/status/1125038440372932608?s=11
42. Bill Murphy, "17 Jeff Bezos Quotes That Suddenly Take on a Whole New Meaning (After 2 Stunning Decisions)," Inc., February 15, 2019, https://www.inc.com/bill-murphy-jr/17-jeff-bezos-quotes-that-suddenly-take-on-a-whole-new-meaning-after-2-stunning-decisions.html
43. Alan Deutschman, "Inside the Mind of Jeff Bezos," Fast Company, August 1, 2004, https://www.fastcompany.com/90435429/this-50-device-is-trying-to-finally-kill-off-the-walkie-talkie
44. Gonzalo Ziadi, "Why You Must Be Stubborn On The Vision, But Flexible On The Details," Medium, April 5, 2018, https://medium.com/the-ascent/why-you-must-be-stubborn-on-the-vision-but-flexible-on-the-details-2b9d86cf20a2
45. Gonzalo Ziadi, "Why You Must Be Stubborn On The Vision, But Flexible On The Details," Medium, April 5, 2018, https://medium.com/the-ascent/why-you-must-be-stubborn-on-the-vision-but-flexible-on-the-details-2b9d86cf20a2
46. "Jeff Bezos live from Code 2016," Recode, May 31, 2016, https://live.recode.net/jeff-bezos-2016-code/
47. Jeff Bezos, "2017 Letter to Shareholders," April 18, 2018, https://blog.aboutamazon.com/company-news/2017-letter-to-shareholders/
48. Steven Levy, "Jeff Bezos Owns the Web in More Ways Than You Think," Wired, November 13, 2011, https://www.wired.com/2011/11/ff_bezos/2

49. Jeff Bezos, "2018 Letter to Shareholders," April 11, 2019, https://blog.aboutamazon.com/company-news/2018-letter-to-shareholders
50. Michael Simmons, "Forget 10,000-Hours: Edison, Bezos and Zuckerberg Follow the 10,000-Experiment Rule," Observer, November, 20, 2017, https://observer.com/2017/11/forget-10000-hours-edison-bezos-zuckerberg-follow-the-10000-experiment-rule/
51. Jon Fortt, "Live Blog: Amazon Unveils Kindle And Bigger Version of the Fire," CNBC, September 6, 2012, https://www.cnbc.com/id/48927497
52. Paul Farhi, "Jeffrey Bezos, Washington Post's next owner, aims for a new 'golden era' at the newspaper," The Washington Post, September 3, 2013, https://www.washingtonpost.com/lifestyle/style/jeffrey-bezos-washington-posts-next-owner-aims-for-a-new-golden-era-at-the-newspaper/2013/09/02/30c00b60-13f6-11e3-b182-1b3bb2eb474c_story.html
53. JP Mangalindan, "Jeff Bezos's mission: Compelling small publishers to think big," Fortune, June 29, 2010, https://fortune.com/2010/06/29/jeff-bezoss-mission-compelling-small-publishers-to-think-big/
54. Charlie Rose, "Amazon's Jeff Bezos looks to the future," CBS News, December 1, 2013, https://www.cbsnews.com/news/amazons-jeff-bezos-looks-to-the-future/
55. Charlie Rose, "Amazon's Jeff Bezos looks to the future," CBS News, December 1, 2013, https://www.cbsnews.com/news/amazons-jeff-bezos-looks-to-the-future/
56. Investor Relations | Amazon.com, "2018 Letter to Shareholders," April 11, 2019, https://ir.aboutamazon.com/static-files/4f64d0cd-12f2-4d6c-952e-bbed15ab1082
57. Charlie Rose, "Amazon's Jeff Bezos looks to the future," CBS News, December 1, 2013, https://www.cbsnews.com/news/amazons-jeff-bezos-looks-to-the-future/
58. Franklin Foer, "Jeff Bezos's Master Plan," The Atlantic, November, 2019,

https://www.theatlantic.com/magazine/archive/2019/11/what-jeff-bezos-wants/598363/

59. "Jeff Bezos live from Code 2016," Recode, May 31, 2016, https://live.recode.net/jeff-bezos-2016-code/

60. Jeff Dyer, Nathan Furr and Mike Hendron, "Leading Like Jeff Bezos Or Elon Musk: Lessons From Their Contrasting Styles," Forbes, September 3, 2019, https://www.forbes.com/sites/nathanfurrjeffdyer/2019/09/03/leading-like-jeff-bezos-or-elon-musk-lessons-from-their-contrasting-styles/#738a399852a7

61. Jessica Stillman, "Jeff Bezos's Motto Is a 2-Word Latin Phrase That Perfectly Captures His Approach to Success," Inc.com, May 30, 2019, https://www.inc.com/jessica-stillman/jeff-bezos-motto-is-a-2-word-latin-phrase-that-perfectly-captures-his-approach-to-success.html

62. Mathias Döpfner, "Jeff Bezos reveals what it's like to build an empire and become the richest man in the world — and why he's willing to spend $1 billion a year to fund the most important mission of his life," Business Insider, April 28, 2018, https://www.businessinsider.com/jeff-bezos-interview-axel-springer-ceo-amazon-trump-blue-origin-family-regulation-washington-post-2018-4

63. Steven Levy, "Jeff Bezos Wants Us All to Leave Earth—for Good," Wired, October 15, 2018, https://www.wired.com/story/jeff-bezos-blue-origin/

64. Adam Lashinsky, "Bezos Prime," Fortune, March 24, 2016, https://fortune.com/longform/amazon-jeff-bezos-prime/

65. Mathias Döpfner, "Jeff Bezos reveals what it's like to build an empire and become the richest man in the world — and why he's willing to spend $1 billion a year to fund the most important mission of his life," Business Insider, April 28, 2018, https://www.businessinsider.com/jeff-bezos-interview-axel-springer-ceo-amazon-trump-blue-origin-family-regulation-washington-post-2018-4

66. Henry Blodget, "I Asked Jeff Bezos The Tough Questions — No Profits, The Book Controversies, The Phone Flop — And He Showed Why Amazon Is Such

A Huge Success," Business Insider, December 13, 2014, https://www.businessinsider.com/amazons-jeff-bezos-on-profits-failure-succession-big-bets-2014-12#ixzz3gdJa8T9C

67. Marina Koren, "Jeff Bezos Has Plans to Extract the Moon's Water," May 10, 2019, https://www.theatlantic.com/science/archive/2019/05/jeff-bezos-moon-nasa/589150/

68. "Jeff Bezos fireside chat at re:MARS 2019", Amazon News, June 6, 2019, https://www.youtube.com/watch?v=AbpXSM8WW4s

69. Franklin Foer, "Jeff Bezos's Master Plan," The Atlantic, November, 2019, https://www.theatlantic.com/magazine/archive/2019/11/what-jeff-bezos-wants/598363/

70. "Jeff Bezos and Caroline Kennedy on plans to head back to the moon," CBS Evening News, July 16, 2019, https://www.youtube.com/watch?v=xEu12Wx2epo&feature=emb_title

71. Steven Levy, "Jeff Bezos Wants Us All to Leave Earth—for Good," Wired, October 15, 2018, https://www.wired.com/story/jeff-bezos-blue-origin/

72. Marina Koren, "Jeff Bezos Has Plans to Extract the Moon's Water," May 10, 2019, https://www.theatlantic.com/science/archive/2019/05/jeff-bezos-moon-nasa/589150/

73. Steven Levy, "Jeff Bezos Owns the Web in More Ways Than You Think," Wired, November 13, 2011, https://www.wired.com/2011/11/ff_bezos/2

74. Dave Mosher, "Jeff Bezos says Amazon is not his 'most important work.' It's this secretive rocket company that toils in the Texas desert," Insider, April 30, 2018, https://www.insider.com/jeff-bezos-blue-origin-rocket-company-most-important-2018-4

75. "Jeff Bezos and Caroline Kennedy on plans to head back to the moon," CBS Evening News, July 16, 2019, https://www.youtube.com/watch?v=xEu12Wx2epo&feature=emb_title

76. Mathias Döpfner, "Jeff Bezos reveals what it's like to build an empire and become the richest man in the world — and why he's willing to spend $1 billion a

year to fund the most important mission of his life," Business Insider, April 28, 2018, https://www.businessinsider.com/jeff-bezos-interview-axel-springer-ceo-amazon-trump-blue-origin-family-regulation-washington-post-2018-4

77. Steven Levy, "Jeff Bezos Wants Us All to Leave Earth—for Good," Wired, October 15, 2018, https://www.wired.com/story/jeff-bezos-blue-origin/

78. Brad Stone, "Jeff Bezos Wants to Deliver You to the Moon," Fortune, May 9, 2019, https://fortune.com/2019/05/09/blue-origin-amazon-jeff-bezos/

79. Steven Levy, "Jeff Bezos Wants Us All to Leave Earth—for Good," Wired, October 15, 2018, https://www.wired.com/story/jeff-bezos-blue-origin/

80. Andrew Nusca, "Amazon CEO Jeff Bezos: We Should Settle Mars 'Because It's Cool'," Fortune, June 1, 2016, https://fortune.com/2016/06/01/amazon-ceo-jeff-bezos-mars/

81. "Jeff Bezos live from Code 2016," Recode, May 31, 2016, https://live.recode.net/jeff-bezos-2016-code/

82. Henry Blodget, "I Asked Jeff Bezos The Tough Questions — No Profits, The Book Controversies, The Phone Flop — And He Showed Why Amazon Is Such A Huge Success," Business Insider, December 13, 2014, https://www.businessinsider.com/amazons-jeff-bezos-on-profits-failure-succession-big-bets-2014-12#ixzz3gdJa8T9C

83. Adam Lashinsky, "Bezos Prime," Fortune, March 24, 2016, https://fortune.com/longform/amazon-jeff-bezos-prime/

84. Adam Lashinsky, "Bezos Prime," Fortune, March 24, 2016, https://fortune.com/longform/amazon-jeff-bezos-prime/

85. Matt Buchanan, "Why Jeff Bezos Bought the Washington Post," The New Yorker, August 5, 2013, https://www.newyorker.com/tech/annals-of-technology/why-jeff-bezos-bought-the-washington-post

86. Paul Farhi, "Jeffrey Bezos, Washington Post's next owner, aims for a new 'golden era' at the newspaper,"

The Washington Post, September 3, 2013, https://www.washingtonpost.com/lifestyle/style/jeffrey-bezos-washington-posts-next-owner-aims-for-a-new-golden-era-at-the-newspaper/2013/09/02/30c00b60-13f6-11e3-b182-1b3bb2eb474c_story.html

87. Paul Farhi, "Jeffrey Bezos, Washington Post's next owner, aims for a new 'golden era' at the newspaper," The Washington Post, September 3, 2013, https://www.washingtonpost.com/lifestyle/style/jeffrey-bezos-washington-posts-next-owner-aims-for-a-new-golden-era-at-the-newspaper/2013/09/02/30c00b60-13f6-11e3-b182-1b3bb2eb474c_story.html
88. Andrew Smith, "Brought to book," The Guardian, February 11, 2001, https://www.theguardian.com/books/2001/feb/11/computingandthenet.technology
89. Steven Mufson, "Bezos courts Washington Post editors, reporters," The Washington Post, September 4, 2013, https://www.washingtonpost.com/business/economy/bezos-courts-washington-post-editors-reporters/2013/09/04/c863516e-156f-11e3-a2ec-b47e45e6f8ef_story.html
90. Steven Mufson, "Bezos courts Washington Post editors, reporters," The Washington Post, September 4, 2013, https://www.washingtonpost.com/business/economy/bezos-courts-washington-post-editors-reporters/2013/09/04/c863516e-156f-11e3-a2ec-b47e45e6f8ef_story.html
91. Henry Blodget, "I Asked Jeff Bezos The Tough Questions — No Profits, The Book Controversies, The Phone Flop — And He Showed Why Amazon Is Such A Huge Success," Business Insider, December 13, 2014, https://www.businessinsider.com/amazons-jeff-bezos-on-profits-failure-succession-big-bets-2014-12#ixzz3gdJa8T9C
92. George Anders, "Jeff Bezos's Top 10 Leadership Lessons," Forbes, April 4, 2012, https://www.forbes.com/sites/georgeanders/2012/04/04/bezos-tips/#4e78e5c72fce

93. Steven Levy, "Jeff Bezos Owns the Web in More Ways Than You Think," Wired, November 13, 2011, https://www.wired.com/2011/11/ff_bezos/2
94. "King of Cyber-Commerce," The Academy of Achievement, May 4, 2001, https://www.achievement.org/achiever/jeffrey-p-bezos/
95. Ivan Minic, "10 Business Lessons From Jeff Bezos," Medium, July 4, 2013, https://medium.com/@burek/10-business-lessons-from-jeff-bezos-3fe167f58d65
96. Erika Andersen, "What Jeff Bezos Knows About Planning vs. Reality," Forbes, September 23, 2013, https://www.forbes.com/sites/erikaandersen/2013/09/23/what-jeff-bezos-knows-about-planning-vs-reality/#604b0bcd2d5b
97. "Top 10 Jeff Bezos Quotes," theDECALIFE, January 12, 2019, https://thedecalife.com/top-10-jeff-bezos-quotes/
98. JP Mangalindan, "Jeff Bezos's mission: Compelling small publishers to think big," Fortune, June 29, 2010, https://fortune.com/2010/06/29/jeff-bezoss-mission-compelling-small-publishers-to-think-big/
99. Natalie Jarvey, "Amazon's Jeff Bezos on Hollywood Strategy: 'When People Join Prime, ... They Buy More Shoes'," The Hollywood Reporter, July 15, 2015, https://www.hollywoodreporter.com/features/amazon-prime-day-ceo-jeff-808535
100. Bill Murphy, "17 Jeff Bezos Quotes That Suddenly Take on a Whole New Meaning (After 2 Stunning Decisions)," Inc., February 15, 2019, https://www.inc.com/bill-murphy-jr/17-jeff-bezos-quotes-that-suddenly-take-on-a-whole-new-meaning-after-2-stunning-decisions.html
101. Jeff Haden, "Billionaire Jeff Bezos: People who are 'right a lot' make decisions differently than everyone else—here's how," CNBC, March 12, 2019, https://www.cnbc.com/2019/03/12/amazon-billionaire-jeff-bezos-explains-why-the-smartest-people-change-their-minds-often.html
102. "Bezos On Innovation," Bloomberg, April 17, 2008, https://www.bloomberg.com/news/articles/2008-04-16/bezos-on-innovation

103. Jessica Stillman, "7 Jeff Bezos Quotes That Outline the Secret to Success," Inc., May 7, 2014, https://www.inc.com/jessica-stillman/7-jeff-bezos-quotes-that-will-make-you-rethink-success.html
104. "The electricity metaphor for the web's future," TED Talks, 2003, https://www.ted.com/talks/jeff_bezos_the_electricity_metaphor_for_the_web_s_future
105. Henry Blodget, "I Asked Jeff Bezos The Tough Questions — No Profits, The Book Controversies, The Phone Flop — And He Showed Why Amazon Is Such A Huge Success," Business Insider, December 13, 2014, https://www.businessinsider.com/amazons-jeff-bezos-on-profits-failure-succession-big-bets-2014-12#ixzz3gdJa8T9C
106. David LaGesse, "America's Best Leaders: Jeff Bezos, Amazon.com CEO," US News, November 19, 2008, https://www.usnews.com/news/best-leaders/articles/2008/11/19/americas-best-leaders-jeff-bezos-amazoncom-ceo
107. Amy Chozick, "How Jeff Bezos Went to Hollywood and Lost Control," The New York Times, March 2, 2019, https://www.nytimes.com/2019/03/02/business/jeff-bezos-lauren-sanchez-amazon-hollywood.html
108. Alan Deutschman, "Inside the Mind of Jeff Bezos," Fast Company, August 1, 2004, https://www.fastcompany.com/50541/inside-mind-jeff-bezos-4
109. Jessica Stillman, "7 Jeff Bezos Quotes That Outline the Secret to Success," Inc., May 7, 2014, https://www.inc.com/jessica-stillman/7-jeff-bezos-quotes-that-will-make-you-rethink-success.html
110. John Greathouse, "5 Time-Tested Success Tips From Amazon Founder Jeff Bezos," Forbes, Apr 30, 2013, https://www.forbes.com/sites/johngreathouse/2013/04/30/5-time-tested-success-tips-from-amazon-founder-jeff-bezos/
111. Alan Deutschman, "Inside the Mind of Jeff Bezos," Fast Company, August 1, 2004, https://www.fastcompany.com/50541/inside-mind-jeff-bezos-4

112. Ilias Tsagklis, “8 Inspiring Jeff Bezos Quotes on Business and Success,” Medium, February 21, 2018, https://medium.com/@iliastsagklis/8-inspiring-jeff-bezos-quotes-on-business-and-success-d22059e51445
113. David LaGesse, “America's Best Leaders: Jeff Bezos, Amazon.com CEO,” US News, November 19, 2008, https://www.usnews.com/news/best-leaders/articles/2008/11/19/americas-best-leaders-jeff-bezos-amazoncom-ceo
114. Adam Lashinsky, “Amazon's Jeff Bezos: The Ultimate Disrupter,” Fortune, November 16, 2012, https://fortune.com/2012/11/16/amazons-jeff-bezos-the-ultimate-disrupter/
115. Alan Deutschman, “Inside the Mind of Jeff Bezos,” Fast Company, August 1, 2004, https://www.fastcompany.com/90435429/this-50-device-is-trying-to-finally-kill-off-the-walkie-talkie
116. “Jeff Bezos nails it on how to succeed in business,” Product Hunt, May 5, 2019, https://twitter.com/producthunt/status/1125038440372932608
117. John Greathouse, “5 (More) Battle-Tested Business Tips From Amazon Founder Jeff Bezos,” Forbes, May 4, 2013, https://www.forbes.com/sites/johngreathouse/2013/05/04/5-more-battle-tested-business-tips-from-amazon-founder-jeff-bezos/
118. Franklin Foer, “Jeff Bezos's Master Plan,” The Atlantic, November, 2019, https://www.theatlantic.com/magazine/archive/2019/11/what-jeff-bezos-wants/598363/
119. George Anders, “Jeff Bezos's Top 10 Leadership Lessons,” Forbes, April 4, 2012, https://www.forbes.com/sites/georgeanders/2012/04/04/bezos-tips/
120. Jeff Bezos, “2018 Letter to Shareholders,” April 11, 2019, https://blog.aboutamazon.com/company-news/2018-letter-to-shareholders
121. “Jeff Bezos fireside chat at re:MARS 2019”, Amazon News, June 6, 2019, https://www.youtube.com/watch?v=AbpXSM8WW4s

122. Mark Sullivan, "Jeff Bezos at re:MARS: 5 standout business tips from the Amazon CEO," Fast Company, June 6, 2019, https://www.fastcompany.com/90360687/jeff-bezos-business-advice-5-tips-from-amazons-remars
123. Siraj Datoo, "How to Succeed in Business, According to Jeff Bezos," Fortune, June 7, 2019, https://fortune.com/2019/06/07/jeff-bezos-tips-advice-entrepreneurs-business/
124. Jeff Bezos, "2018 Letter to Shareholders," April 11, 2019, https://blog.aboutamazon.com/company-news/2018-letter-to-shareholders
125. Natalie Jarvey, "Amazon's Jeff Bezos on Hollywood Strategy: 'When People Join Prime, ... They Buy More Shoes'," The Hollywood Reporter, July 15, 2015, https://www.hollywoodreporter.com/features/amazon-prime-day-ceo-jeff-808535
126. Rose Leadem, "10 Inspiring Quotes From the Fearless Jeff Bezos," Entrepreneur Magazine, May 24, 2017, https://www.entrepreneur.com/slideshow/294490
127. "Top 10 Jeff Bezos Quotes," theDECALIFE, January 12, 2019, https://thedecalife.com/top-10-jeff-bezos-quotes/
128. Steven Levy, "Jeff Bezos Owns the Web in More Ways Than You Think," Wired, November 13, 2011, https://www.wired.com/2011/11/ff_bezos/2
129. Steven Levy, "Jeff Bezos Owns the Web in More Ways Than You Think," Wired, November 13, 2011, https://www.wired.com/2011/11/ff_bezos/2
130. Ivan Minic, "10 Business Lessons From Jeff Bezos," Medium, July 4, 2013, https://medium.com/@burek/10-business-lessons-from-jeff-bezos-3fe167f58d65
131. Jessica Stillman, "7 Jeff Bezos Quotes That Outline the Secret to Success," Inc., May 7, 2014, https://www.inc.com/jessica-stillman/7-jeff-bezos-quotes-that-will-make-you-rethink-success.html
132. Investor Relations | Amazon.com, "2012 Letter to Shareholders," April, 2013 https://ir.aboutamazon.com/static-files/c262ba80-84d1-4fa5-a1bd-fde2f7d706d5

133. Ivan Minic, "10 Business Lessons From Jeff Bezos," Medium, July 4, 2013, https://medium.com/@burek/10-business-lessons-from-jeff-bezos-3fe167f58d65
134. Henry Blodget, "I Asked Jeff Bezos The Tough Questions — No Profits, The Book Controversies, The Phone Flop — And He Showed Why Amazon Is Such A Huge Success," Business Insider, December 13, 2014, https://www.businessinsider.com/amazons-jeff-bezos-on-profits-failure-succession-big-bets-2014-12
135. Charlie Rose, "Amazon's Jeff Bezos looks to the future," CBS News, December 1, 2013, https://www.cbsnews.com/news/amazons-jeff-bezos-looks-to-the-future/
136. Ivan Minic, "10 Business Lessons From Jeff Bezos," Medium, July 4, 2013, https://medium.com/@burek/10-business-lessons-from-jeff-bezos-3fe167f58d65
137. "The List: CNBC First 25," CNBC, April 29, 2014, https://www.cnbc.com/2014/04/29/cnbc-25-jeff-bezos.html
138. George Anders, "Jeff Bezos's Top 10 Leadership Lessons," Forbes, April 4, 2012, https://www.forbes.com/sites/georgeanders/2012/04/04/bezos-tips/
139. Brad Stone, "AI, Trump, and Gawker: Six Highlights From Amazon's Jeff Bezos Interview," Bloomberg, June 1, 2016, https://www.bloomberg.com/news/articles/2016-06-01/ai-trump-and-gawker-six-highlights-from-amazon-s-jeff-bezos-interview
140. John Cook, "Jeff Bezos on innovation: Amazon 'willing to be misunderstood for long periods of time'," GeekWire, June 7, 2011, https://www.geekwire.com/2011/amazons-bezos-innovation/
141. "Jeff Bezos live from Code 2016," Recode, May 31, 2016, https://live.recode.net/jeff-bezos-2016-code/
142. "Top 10 Jeff Bezos Quotes," theDECALIFE, January 12, 2019, https://thedecalife.com/top-10-jeff-bezos-quotes/

143. Mathias Döpfner, "Jeff Bezos reveals what it's like to build an empire and become the richest man in the world — and why he's willing to spend $1 billion a year to fund the most important mission of his life," Business Insider, April 28, 2018, https://www.businessinsider.com/jeff-bezos-interview-axel-springer-ceo-amazon-trump-blue-origin-family-regulation-washington-post-2018-4
144. George Anders, "Jeff Bezos's Top 10 Leadership Lessons," Forbes, April 4, 2012, https://www.forbes.com/sites/georgeanders/2012/04/04/bezos-tips/
145. Steven Levy, "Jeff Bezos Owns the Web in More Ways Than You Think," Wired, November 13, 2011, https://www.wired.com/2011/11/ff_bezos/2
146. David LaGesse, "America's Best Leaders: Jeff Bezos, Amazon.com CEO," US News, November 19, 2008, https://www.usnews.com/news/best-leaders/articles/2008/11/19/americas-best-leaders-jeff-bezos-amazoncom-ceo
147. Ivan Minic, "10 Business Lessons From Jeff Bezos," Medium, July 4, 2013, https://medium.com/@burek/10-business-lessons-from-jeff-bezos-3fe167f58d65
148. "Jeff Bezos," Esquire, September 25, 2008, https://www.esquire.com/news-politics/a5038/jeff-bezos-1008/
149. Ivan Minic, "10 Business Lessons From Jeff Bezos," Medium, July 4, 2013, https://medium.com/@burek/10-business-lessons-from-jeff-bezos-3fe167f58d65
150. Catherine Clifford, "Jeff Bezos: 'If you cannot afford to be misunderstood, don't do anything new or innovative'," CNBC, May 17, 2018, https://www.cnbc.com/2018/05/17/jeff-bezos-on-what-it-takes-to-be-innovative.html
151. Ivan Minic, "10 Business Lessons From Jeff Bezos," Medium, July 4, 2013, https://medium.com/@burek/10-business-lessons-from-jeff-bezos-3fe167f58d65
152. Ilias Tsagklis, "8 Inspiring Jeff Bezos Quotes on Business and Success," Medium, February 21, 2018, https://medium.com/@iliastsagklis/8-inspiring-jeff-

bezos-quotes-on-business-and-success-d22059e51445
153. Mathias Döpfner, "Jeff Bezos reveals what it's like to build an empire and become the richest man in the world — and why he's willing to spend $1 billion a year to fund the most important mission of his life," Business Insider, April 28, 2018, https://www.businessinsider.com/jeff-bezos-interview-axel-springer-ceo-amazon-trump-blue-origin-family-regulation-washington-post-2018-4
154. John Greathouse, "5 Time-Tested Success Tips From Amazon Founder Jeff Bezos," Forbes, Apr 30, 2013, https://www.forbes.com/sites/johngreathouse/2013/04/30/5-time-tested-success-tips-from-amazon-founder-jeff-bezos/
155. Alan Deutschman, "Inside the Mind of Jeff Bezos," Fast Company, August 1, 2004, https://www.fastcompany.com/50541/inside-mind-jeff-bezos-4
156. Charlie Rose, "Amazon's Jeff Bezos looks to the future," CBS News, December 1, 2013, https://www.cbsnews.com/news/amazons-jeff-bezos-looks-to-the-future/
157. Adam Lashinsky, "Amazon's Jeff Bezos: The Ultimate Disrupter," Fortune, November 16, 2012, https://fortune.com/2012/11/16/amazons-jeff-bezos-the-ultimate-disrupter/
158. Steven Levy, "Jeff Bezos Owns the Web in More Ways Than You Think," Wired, November 13, 2011, https://www.wired.com/2011/11/ff_bezos/2
159. "Profile: Jeff Bezos," BBC, August 6, 2013, https://www.bbc.com/news/business-23587389
160. Steven Levy, "Jeff Bezos Owns the Web in More Ways Than You Think," Wired, November 13, 2011, https://www.wired.com/2011/11/ff_bezos/2
161. Rose Leadem, "10 Inspiring Quotes From the Fearless Jeff Bezos," Entrepreneur Magazine, May 24, 2017, https://www.entrepreneur.com/slideshow/294490
162. Steven Mufson, "Bezos courts Washington Post editors, reporters," The Washington Post, September 4, 2013, https://www.washingtonpost.com/business/econom

y/bezos-courts-washington-post-editors-reporters/2013/09/04/c863516e-156f-11e3-a2ec-b47e45e6f8ef_story.html
163. Steven Levy, "Jeff Bezos Owns the Web in More Ways Than You Think," Wired, November 13, 2011, https://www.wired.com/2011/11/ff_bezos/2
164. "Jeff Bezos," Esquire, September 25, 2008, https://www.esquire.com/news-politics/a5038/jeff-bezos-1008/
165. Jeff Haden, "20 Years Ago, Jeff Bezos Said This 1 Thing Separates People Who Achieve Lasting Success From Those Who Don't," Inc., November 6, 2017, https://www.inc.com/jeff-haden/20-years-ago-jeff-bezos-said-this-1-thing-separates-people-who-achieve-lasting-success-from-those-who-dont.html
166. "Jeff Bezos nails it on how to succeed in business," Product Hunt, May 5, 2019, https://twitter.com/producthunt/status/1125038440372932608
167. "Top 10 Jeff Bezos Quotes," theDECALIFE, January 12, 2019, https://thedecalife.com/top-10-jeff-bezos-quotes/
168. Brian Morrissey, "Jeff Bezos, Amazon.com," Adweek, September 14, 2009, https://www.adweek.com/brand-marketing/jeff-bezos-amazoncom-94441/
169. Jessica Stillman, "7 Jeff Bezos Quotes That Outline the Secret to Success," Inc., May 7, 2014, https://www.inc.com/jessica-stillman/7-jeff-bezos-quotes-that-will-make-you-rethink-success.html
170. Bill Murphy, "17 Jeff Bezos Quotes That Suddenly Take on a Whole New Meaning (After 2 Stunning Decisions)," Inc., February 15, 2019, https://www.inc.com/bill-murphy-jr/17-jeff-bezos-quotes-that-suddenly-take-on-a-whole-new-meaning-after-2-stunning-decisions.html
171. "Online Extra: Jeff Bezos on Word-of-Mouth Power", Bloomberg, August 2, 2004, https://www.bloomberg.com/news/articles/2004-08-01/online-extra-jeff-bezos-on-word-of-mouth-power
172. Cal Fussman, "Jeff Bezos: What I've Learned," Esquire, January, 2002,

https://www.esquire.com/news-politics/interviews/a2033/esq0102-jan-bezos/

173. Alan Deutschman, "Inside the Mind of Jeff Bezos," Fast Company, August 1, 2004, https://www.fastcompany.com/50541/inside-mind-jeff-bezos-4

174. George Anders, "Jeff Bezos's Top 10 Leadership Lessons," Forbes, April 4, 2012, https://www.forbes.com/sites/georgeanders/2012/04/04/bezos-tips/

175. Franklin Foer, "Jeff Bezos's Master Plan," The Atlantic, November, 2019, https://www.theatlantic.com/magazine/archive/2019/11/what-jeff-bezos-wants/598363/

176. Jodi Kantor and David Streitfeld, "Inside Amazon: Wrestling Big Ideas in a Bruising Workplace," The New York Times, August 15, 2015, https://www.nytimes.com/2015/08/16/technology/inside-amazon-wrestling-big-ideas-in-a-bruising-workplace.html

177. Adam Lashinsky, "Amazon's Jeff Bezos: The Ultimate Disrupter," Fortune, November 16, 2012, https://fortune.com/2012/11/16/amazons-jeff-bezos-the-ultimate-disrupter/

178. David LaGesse, "America's Best Leaders: Jeff Bezos, Amazon.com CEO," US News, November 19, 2008, https://www.usnews.com/news/best-leaders/articles/2008/11/19/americas-best-leaders-jeff-bezos-amazoncom-ceo

179. Jon Fortt, "Live Blog: Amazon Unveils Kindle And Bigger Version of the Fire," CNBC, September 6, 2012, https://www.cnbc.com/id/48927497

180. "Jeff Bezos," Esquire, September 25, 2008, https://www.esquire.com/lifestyle/money/news/a55701/jeff-bezos-amazon-interview-1997/

181. Jon Fortt, "Live Blog: Amazon Unveils Kindle And Bigger Version of the Fire," CNBC, September 6, 2012, https://www.cnbc.com/id/48927497

182. Henry Blodget, "I Asked Jeff Bezos The Tough Questions — No Profits, The Book Controversies, The Phone Flop — And He Showed Why Amazon Is Such A Huge Success," Business Insider, December 13, 2014, https://www.businessinsider.com/amazons-

jeff-bezos-on-profits-failure-succession-big-bets-2014-12
183. Steven Levy, "Jeff Bezos Owns the Web in More Ways Than You Think," Wired, November 13, 2011, https://www.wired.com/2011/11/ff_bezos/2
184. "The Inner Bezos," March 01, 1999, https://www.wired.com/1999/03/bezos-3/
185. Charlie Rose, "Amazon's Jeff Bezos looks to the future," CBS News, December 1, 2013, https://www.cbsnews.com/news/amazons-jeff-bezos-looks-to-the-future/
186. George Anders, "Jeff Bezos's Top 10 Leadership Lessons," Forbes, April 4, 2012, https://www.forbes.com/sites/georgeanders/2012/04/04/bezos-tips/
187. "Jeff Bezos live from Code 2016," Recode, May 31, 2016, https://live.recode.net/jeff-bezos-2016-code/
188. "Jeff Bezos fireside chat at re:MARS 2019", Amazon News, June 6, 2019, https://www.youtube.com/watch?v=AbpXSM8WW4s
189. Mathias Döpfner, "Jeff Bezos reveals what it's like to build an empire and become the richest man in the world — and why he's willing to spend $1 billion a year to fund the most important mission of his life," Business Insider, April 28, 2018, https://www.businessinsider.com/jeff-bezos-interview-axel-springer-ceo-amazon-trump-blue-origin-family-regulation-washington-post-2018-4
190. "The electricity metaphor for the web's future," TED Talks, 2003, https://www.ted.com/talks/jeff_bezos_the_electricity_metaphor_for_the_web_s_future
191. Cal Fussman, "Jeff Bezos: What I've Learned," Esquire, January, 2002, https://www.esquire.com/news-politics/interviews/a2033/esq0102-jan-bezos/
192. Cal Fussman, "Jeff Bezos: What I've Learned," Esquire, January, 2002, https://www.esquire.com/news-politics/interviews/a2033/esq0102-jan-bezos/
193. Rose Leadem, "10 Inspiring Quotes From the Fearless Jeff Bezos," Entrepreneur Magazine, May

24, 2017,
https://www.entrepreneur.com/slideshow/294490

194. Bryan Collins, "Jeff Bezos Says Successful People Make These Two Types Of Decisions," Forbes, May 7, 2019, https://www.forbes.com/sites/bryancollinseurope/2019/03/07/jeff-bezos-says-successful-people-make-these-two-types-of-decisions/
195. "Jeff Bezos live from Code 2016," Recode, May 31, 2016, https://live.recode.net/jeff-bezos-2016-code/
196. Rose Leadem, "10 Inspiring Quotes From the Fearless Jeff Bezos," Entrepreneur Magazine, May 24, 2017, https://www.entrepreneur.com/slideshow/294490
197. Jessica Stillman, "7 Jeff Bezos Quotes That Outline the Secret to Success," Inc., May 7, 2014, https://www.inc.com/jessica-stillman/7-jeff-bezos-quotes-that-will-make-you-rethink-success.html
198. Steven Mufson, "Bezos courts Washington Post editors, reporters," The Washington Post, September 4, 2013, https://www.washingtonpost.com/business/economy/bezos-courts-washington-post-editors-reporters/2013/09/04/c863516e-156f-11e3-a2ec-b47e45e6f8ef_story.html
199. David LaGesse, "America's Best Leaders: Jeff Bezos, Amazon.com CEO," US News, November 19, 2008, https://www.usnews.com/news/best-leaders/articles/2008/11/19/americas-best-leaders-jeff-bezos-amazoncom-ceo

Follow me on social media:

https://twitter.com/olivia_longray
https://facebook.com/olivia.longray
https://instagram.com/olivia_longray

Cover photo by Steve Jurvetson on Flickr

Printed in Great Britain
by Amazon